Naiara Sampaio
Domingos L. dos S. Silva
Gonçalo M. da Conceição

CURNERACEAE Kunth ex DC

Naiara Sampaio
Domingos L. dos S. Silva
Gonçalo M. da Conceição

CURNERACEAE Kunth ex DC

ScienciaScripts

Imprint
Any brand names and product names mentioned in this book are subject to trademark, brand or patent protection and are trademarks or registered trademarks of their respective holders. The use of brand names, product names, common names, trade names, product descriptions etc. even without a particular marking in this work is in no way to be construed to mean that such names may be regarded as unrestricted in respect of trademark and brand protection legislation and could thus be used by anyone.

Cover image: www.ingimage.com

This book is a translation from the original published under ISBN 978-3-330-74690-9.

Publisher:
Sciencia Scripts
is a trademark of
Dodo Books Indian Ocean Ltd. and OmniScriptum S.R.L publishing group

120 High Road, East Finchley, London, N2 9ED, United Kingdom
Str. Armeneasca 28/1, office 1, Chisinau MD-2012, Republic of Moldova, Europe
Printed at: see last page
ISBN: 978-620-8-08407-3

ACKNOWLEDGEMENTS

Finally, another stage is coming to an end and so new stages are beginning in a cycle full of benefits and improvements acquired along a journey full of situations, both sad and happy, that we have been through and, above all, we have met people who have been important in our lives and who, in one way or another, could not fail to be mentioned in my trajectory of achievements.

I am infinitely grateful to my almighty father, creator of heaven and earth, to our master **God**, who has given me the opportunity to grow and fulfil myself at university and in everything that has been possible for me throughout this stage of my degree. Thank you, Father, for your unconditional love and for making me mature and grow in my ideas and purpose in life alongside my friends. During this time, I learnt to see life as a puzzle, and that in each piece there is a realisation.

To Dr **Gonçalo Mendes da Conceição**, my teacher, friend, father, brother and finally my advisor, for his trust and enormous patience, for passing on his knowledge to me from the moment I entered the Plant Biology laboratory at CESC/UEMA, making me grow intellectually and improving my knowledge. Thank you so much, Professor, for the opportunities you've given me in my professional career and a thousand apologies for the mistakes I've made, without you I wouldn't be where I am now. I must also not forget your family (wife, children, grandchildren and even sons-in-law and daughters-in-law), for their enormous understanding in your absence and for some of them accompanying you to university so that you could support your students. Thank you all so much!

I would like to thank my parents **Luiz Machado Sampaio and Natal de Maria Assunção Sampaio**, who together were fundamental in my graduation, because through them I began this phase and did not have to stop my studies at any time, and they passed on to me their best inheritance and characteristic: Education. Thank you mum for being my son's mother when I was away from him, thanks to my mum I was able to take part in events and congresses held during university, and thank you dad for guiding me in my determinations, I just have to tell you that you are my reason for living, I LOVE YOU.

It would be impossible not to thank my husband **Laerte de Almada Sá**, because I know I was a pain in the arse during my training and yet he stayed by my side and put up with me throughout my professional career. I was able to grow in the writing of my monograph thanks to his great love and affection. This was fundamental to the development and finalisation of this study through your understanding and enormous patience. Thank you so much for coming into my life. And to my son **Kayllon Sampaio Sá,** because even with his innocence he gave me the strength to go on and the affection to grow, I love you unconditionally my baby.

To my siblings **Clairton Assunção Sampaio and Laide Assunção Sampaio and** to my step-sisters **Gleice de Castro Assunção and Emilly Rayane Assunção Silva**, for giving me strength and helping me in my education and for being part of my life, thank you for everything, I've learnt a lot from you.

To all my family and friends, who even if they were far away or absent at times, wished me success on this long journey, especially my aunts **Edna Maria, Maria de Conceição and Rozana Ferreira** and my friend and sister **Aline Gomes,** my brothers-in-law **Layo de**

Almada, Laylson de Almada, Fabrícia de Almada and Daniele de Almada, and my parents-in-law **Pedro Gomes and Lucilene Sá.** Without you in my life, I wouldn't have made it to the end of writing this final project, thank you for being part of my life.

During the process of writing the descriptions of the Turneraceae species, I would like to thank the collaboration of professors **Me. Eduardo Silva Oliveira and Me. Lamarck do Nascimento Galdino da Rocha,** for all their help and assistance, thank you for all their teaching.

Whilst at university, I would like to thank my friends from undergraduate course 2010.1, **Juliete Lima, Cintia Fernandes, Lucianna Lima, Francisca Natália, Ricardo Oliveira, Maria dos Remédios, Stênio Raniery, Rosa Cristina, Maxcilene da Silva, Silmara Gomes, Cleyton Ferreira, Milane Oliveira, Gracyone Silva, Carmem Célia, Domingos Lucas, Liliane Torres and Larissa Fernandes**, who shared special moments with me throughout my time at UEMA.

To my lab friends **Daniele Souza, Paula Regina, Filipe Bezerra, Domingos Lucas, Débora Andrade, Dailma Medeiros, Ingrid Abreu, Lucianna Lima, Silmara Gomes, Werton Nobre, Adriana Oliveira, Cleyton Ferreira, Maycon Adams, Maria da Conceição, Gizele Brito, Cristhian Franco, Guilherme Silva and Jadson Lima,** because the time we spent together made me acquire enriching knowledge for my learning. Among the members of LABIVE, I would especially like to thank **Débora, Adriana, Daniele Lucas and Maycon** for all their help in collecting and herborising the botanical material. As well as being friends, you were brothers and sisters and helped me through the difficult times of writing the monograph.

Francisco de Paula Athayde Filho, my short-term internship supervisor, for his enormous help in taxonomy, even though the internship was in Cryptogams, he collaborated a lot in carrying out this research, even though he didn't follow the research in Pteridophytes, but he was of enormous help in the field work, in the description and identification of the Turneraceae species. To my internship friends at the Cryptogams Laboratory at UNEMAT (Nova Xavantina/MT), Professor **Carlos Kreutz,** and students **Josiene**, **Sandra, César and Miquéia**, for being important in helping me with my internship and making my days more fun and happy.

The **State University of Maranhão** for my academic training and to all the staff, especially **Seu Raimundo, Tia Jura**, **Tia Antônia**, **Cleyton, André,** and **Gil**, where they were present during my time at UEMA and during my academic research.

And to all my Biological Sciences teachers, who were part of my education and made my learning easier, making me a better professional. I would also like to thank the examining board for agreeing to share this very special moment in my life.

I **would** like to thank the Maranhão Research and Scientific Development Foundation - **FAPEMA,** for all the funding I received, especially for the short-term internship in Cryptogams, which helped me a lot with my taxonomy.

And why not mention the friendly and hospitable people who made the botanical collections more enjoyable and with a taste of coming back again, I am eternally grateful to the residents of the **Buriti do Meio Settlement**, especially **Tia Helena and Seu Matias, Geraldo and Francisco de Assis,** for their accommodation,

accompaniment and help with the collections.
And to everyone who has contributed in one way or another to this dream come true, and to you too, the **reader**, thank you so much!!!

Naiara Assunção Sampaio

PRESENTATION

The book deals with the taxonomic aspects of the Turneraceae species occurring in a Cerrado fragment in the municipality of Caxias, Maranhão, Brazil. A species list, identification key and description of the seven species of Turneraceae occurring in the region, geographical distribution, common name, material examined and the main morphological aspects for identifying the species are presented.

As this is a botanical family of great economic and environmental importance, and because this book presents new information on this plant group, it is considered that this work will serve all those interested in the knowledge, utilisation and conservation of the biodiversity of Brazil's Cerrado areas.

We therefore recommend that you read and use the information contained in this book dedicated to one of the families of Brazilian flora.

The authors

1. INTRODUCTION

Taxonomy and Plant Systematics deal with the need to understand the identification of biological elements and to explain the origin of species, where the aim is to study the diversity of botanical organisms, organising their internal and/or external morphological characteristics, thus understanding the identification of species (CRONQUIST, 1988; WEBERLING; SCHWANTES, 1986).

Taxonomic or systematic identification points to the determination of a taxon as identical/similar to another that already exists and can be used to confirm it. This can be done by comparison with duly identified herbarium material, by the use of dichotomous identification keys and by specific literature, which during the identification of specimens can describe new taxa for science, as recommended by the International Code of Botanical Nomenclature - CINB (SUBRAHMANYAM, 1995).

Botanical identification and plant morphology are also fundamental to taxonomic knowledge, but taxonomy does not only require morphological characteristics, but is also based on other areas such as anatomy and genetics (HARLOW et al., 1991).

With its climatic diversity, Brazil has a wide range of floral species distributed throughout its vast territory, and it is known that many of these plant species have not yet been discovered or are yet to be identified (KAMPF, 2000; MARX, 2004).

Turneraceae Kunth ex DC. has around 12 genera and 226 species, distributed throughout the tropical and subtropical regions of the world, with America and Africa as its centre of diversity (ARBO, 2014; THULIN et al., 2012). In Brazil, there are approximately 156 species and two genera, *Turnera* and *Piriqueta*, however, *Turnera* is the most representative, with around 119 species, followed by *Piriqueta* with 37 species, both found in typical cerrado and rupestrian grassland environments (ARBO, 2014), and among these, 16 species are referenced for Maranhão by the Lista da Flora do Brasil (2017).

The Turneraceae family includes shrubs, subshrubs and herbs with alternate, simple, entire or lobed leaves, with glands on the petiole or at the base of the blade; with small stipules or absent; androgynous, regular and actinomorphic flowers, characterised by having five sepals inserted in the calyx, deciduous and generally partially united, forming a campanulate or cylindrical tube; the stamens are free, the flowers have a supernate, unilocular or tricarpellar ovary; the fruits are capsular, globose (BRITO FILHO, 2011).

As mentioned in the work by Mendonça et al. (2007), entitled vascular flora of the cerrado biome, a total of 6,671 native *taxa* were recorded, distributed in 170 families and 1,144 genera. Among these families, Turneraceae is represented with two genera, *Piriqueta* Aubl., with 25 species and *Turnera* L. with 21 species, totalling 46 species. In the work by Nascimento; Conceição (2011), the species *Turnera ulmifolia* L. is indicated for therapeutic purposes such as kidney stones, liver, with the root being used as the plant part and decoct or infusion as the form of preparation.

Studies on Turneraceae in Maranhão are still scarce, with only two species catalogued, in the work by Conceição et al. (2012) entitled "Florula fanerogâmica da Área de Proteção Ambiental Municipal do Inhamum, Caxias/MA, Brasil" In: Biodiversidade na Área de Proteção Ambiental do Inhamum, the species of the genus

Turnera - *Turnera* sp. are mentioned. and *Turnera ulmifolia* L. The latter was also found in the Olho d'água do Raposo community in the same municipality by Nascimento; Conceição (2011).

Maranhão has a diverse flora, which makes it important to study plant taxonomy, favouring areas of learning, as a fundamental science capable of subsidising applied knowledge in botany. This pioneering study of the Turneraceae family makes it possible to give this cosmopolitan family a taxonomic overview, with the *Turnera* genus being the most representative of the two found in Brazil. This same genus is very important in the Brazilian economy as it is used for ornamental and medicinal purposes.

The species of Turneraceae in Maranhão still need further study in order to know and determine the main species for the area under study, as well as to contribute to providing data for the state. Studies and research should therefore be carried out in order to learn more about the species of Turneraceae occurring in the municipality of Caxias and consequently expand the taxonomic knowledge of this important family for Maranhão.

1.1 TAXONOMY

The international standard ANSI/NISO (2005) defines a taxonomy as a vocabulary consisting of terms that are preferred by the organisation, or a collection of language terms that are hierarchically grouped to make them easier to identify.

The concern to establish the different groups of organisms is still a very old classification system, dating back to pre-Christian and pre-Aristotelian civilisations, with Luca Ghini, Césalpin and the

Bauhin brothers, where scientific plant systematics began, but it was only at the end of the 17th century and the beginning of the 18th century that an attempt was made to develop a method for describing and classifying plants (SAMPAIO, 2006).

Sampaio (2006) points out that according to the taxonomic classification used by John Ray in 1692, the criterion for classification was no longer the type of habitat of the organism, but the similarity of its reproductive organs. And it was only Linnaeus (1707-1788) who founded modern plant systematics and taxonomy. In 1737, he introduced a classification system based on the characters of reproductive structures, creating a hierarchy of five taxonomic categories.

The classification rules of plant taxonomy have opened the door to a new period, in relation to phylogenetic studies and their correlation of morphological characters, evolving plants. These rules are based on the identification of all species of flora, grouping them by morphological similarity and by genetic clade, which is characterised by allowing a structure in the principles of botanical kinship (BRASIL, 2009).

1.2 THE FAMILY TURNERACEAE Kunth ex DC.

The list of the flora of Brazil 2020 records the occurrence of two genera in the Turneraceae family: *Piriqueta* and *Turnera*. The latter genus is the most representative of this family. According to Brito Filho (2011), these genera are found in the Rupestrian Fields and Cerrado.

Some floristic and taxonomic studies have contributed to the systematics of Turneraceae, including the works of Arbo (1995, 1997, 2000, 2005, 2008); Batista (2005); Camargo; Vilegas (2008); Rocha et al. (2012); Urban (1883). Urban's (1883) monograph, for example, describes 54 species of the genus *Turnera* distributed in nine series: *Salicifoliae, Stenodictyae, Anomalae, Leiocarpae, Annulares, Papilliferae, Microphyllae, Capitatae* and *Turnera.* Following this division, Arbo (2008) also indicated two other series, *Conciliatae* and *Sessilifoliae*.

Stevens (2001) proposes that the Turneraceae and Malesherbiaceae families are linked to the Passifloraceae *sensu lato* clade, with which they share several common characters, among which Zamberlan (2007) highlights the presence of cyanogenic glycosides, leaf glands and plastidial inheritance for some species, which form a paraphyletic group in Malpighiales, according to the Angiosperm Phylogeny Group III (APG III, 2009).

The family has medicinal importance and was used in Mexico and Cuba, where the Indians used the aqueous extract of the whole *Turnera diffusa* plant as an expectorant, diuretic, aphrodisiac and in the treatment of spermatorrhoea, otitis and nephritis. This is probably because the Turneraceae family contains fatty acids, flavonoids and alkaloids (ANTÔNIO, 1996; BRITO FILHO, 2011).

1.2.1 TAXONOMIC CLASSIFICATION

Plant **Kingdom**

Division Magnoliophyta

Class Magnoliopsida

Order Malpighiales

Family Turneraceae

1.3 THE GENUS *Turnera* L.

Turnera species are herbaceous and shrubby, with simple leaves, with or without stipules and sometimes reduced to collettes, serrate and reticulate margins, often with nectary glands and trichomes, with a pedicel attached totally or partially to the petiole; the flowers are heterostylous or homostylous, with an undeveloped pedicel, and a corolla with white, yellow or orange petals, which may or may not be maculate at the base;the staminal filaments are attached to the base of the calyx; the ovary can vary from ovoid to globose, with filiform stipes, fimbriated to branched stigmas; the fruit is a warty, smooth, tomentose, pubescent or puberulent capsule; the seeds are reticulate, glabrous or papillose, with a unilateral or enveloping aril (ROCHA et al., 2012).

According to Neffa; Fernández (2000), the *Turnera* genus is one of the most important in Brazil and its species are commonly found in open fields and along roadsides. Some species of the *Turnera* genus are used for medicinal purposes, highlighting the studies by Antônio (1996); Antônio; Souza (1998); Barbosa et al. (2007) and Nascimento; Conceição (2011).

1.3.1 geographic distribution of the genus *turnera* in brazil

The *Turnera* genus in Brazil has approximately 119 species, distributed in all Brazilian regions (North, Northeast, South, Southeast and Centre-West), with the highest number of occurrences in the Cerrado phytogeographic domain (72 species), Caatinga (34), Atlantic Forest (33), Amazon (28), Pantanal (06) and Pampa (1), according to the Flora of Brazil 2020 species list (Figure 1).

Fig 1. Area of occurrence of the genus *Turnera* and number of species per state.

1.4 THE GENDER *Piriqueta* Aulb.

Piriqueta species have a sub-shrub or herbaceous perennial habit; rarely annuals; little branched; the trichomes are setiform glandular with dilated bases and/or simple stellate detectors; their leaves have no stipules; nectaries are absent; they have uniflorous, axillary, cymose inflorescences or gathered in terminal racemes; they have a developed peduncle, usually without bracts and bracteoles; the flowers can be homostylous or heterostylous with a developed pedicel; the corolla varies from white, yellow, salmon to pink, with obovate petals, a fimbriate membranaceous corona, inserted at the base of the petal blade and under the sepals; the ovary is tomentose; filiform stipes; fimbriate stigmas; they have strigose, smooth or warty capsules; and obovoid seeds, unilateral aril (ROCHA et al., 2012).

According to Silva et al. (2013), the genus *Piriqueta* is made up of distichous species. Rocha et al. (2012) emphasises that this occurs in open environments, especially in places disturbed by anthropogenic action. Studies on this genus can be found in the works by Arbo (1995) and Silva et al. (2013).

1.4.1 GEOGRAPHICAL DISTRIBUTION OF THE GENDER *Piriqueta* IN BRAZIL

According to the Flora of Brazil 2020 species list, the genus Piriqueta is widely distributed in all regions of Brazil (North, Northeast, Centre-West, Southeast, South), as well as in all of the country's phytogeographic domains (Amazon, Caatinga, Cerrado, Atlantic Forest, Pampa and Pantanal) (Figure 2).

Fig 2. Occurrence of the genus *Piríqueta* and number of species per state.

2. BURITI DO MEIO ENVIRONMENTAL PROTECTION AREA

The Buriti do Meio Environmental Protection Area (APA) is located in the municipality of Caxias/MA, created by Law No. 1.540/2004 of 25 March 2004, in the Buriti do Meio and Santa Rosa Settlement Project, part of the 2nd District of Caxias/MA. It has an area of 58,347.30 ha, about 35 km from the urban perimeter, with geographical coordinates -04° 54' 48.1" South latitude and -043° 06' 49.2" West longitude (Figure 3). The vegetation that makes up the APA is predominantly Cerrado, but there is also riparian forest or gallery forest along the streams, lakes and springs. The hydrology of the area is mainly characterised by the Buriti do Meio and Riachão streams (Figure 4) (BRASIL, 2006).

Fig. 3 Map of the Buriti do Meio Environmental Protection Area. Source: Google Earth Pro 2016; ArcGis 9.3. Organisation: SILVA, W. F. N. 2016.

Hydrographic Map and Location of the Buriti do Meio Municipal APA, Caxias - MA,

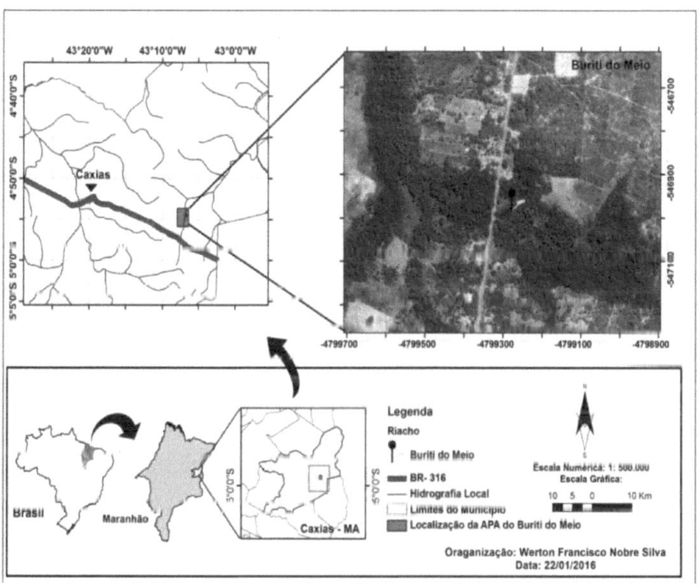

Fig. 4 (A, B, C and D) - General view of the study area in the Buriti do Meio APA, Caxias/MA. Photo: SAMPAIO, N. A.

3. COLLECTION AND HERBORISATION OF BOTANICAL MATERIAL

Specimens were collected from 3-5 samples of each species, when possible, in their reproductive period (with flowers, fruits and/or flower buds) to help with the identification and description of morphological characters. The specimens were collected from roadsides, open environments and trails within the vegetation, with random walks around the study area every fortnight (Figure 5).

The specimens collected were photographed and placed in

wooden presses interspersed with newspaper and cardboard, while still in the field. After pressing, the botanical material was taken to the Plant Biology Laboratory (LABIVE) at the Caxias Higher Education Centre of the State University of Maranhão (CESC/UEMA) for drying at room temperature. After drying the botanical material, the exsiccates were prepared using wood paper and paper 40 in the standard size of the Prof. Aluízio Bittencourt Herbarium (HABIT) and duly labelled (Figure 5 A and B).

4. TAXONOMY AND IDENTIFICATION OF BOTANICAL MATERIAL

At the end of the botanical herborisation procedure, each sample collected was studied using a Stemi DV4 Carl Zeiss microscope to describe the specimens after analysing their morphological characteristics, such as: shape, consistency, hairiness and colour of the stem; length, shape, margin and presence of glands on the stipules; shape, phyllotaxis, length, width, apex, base, margin of the leaves; length and hairiness of the peduncle; shape, length, width, colour, margin, hairiness, presence of glands on the sepals; shape, length, number of glands, hairiness and margin of the petals; type, shape, length and width of the fruit; shape, length, hairiness of the seeds; presence of glands defining which structure is present, colouring of the calyx and corolla, among other characteristics, with the aim of developing a taxonomic key that better identifies the species inventoried, prioritising more specific characteristics for each species.

Taxonomic identification was carried out at the Plant Biology Laboratory - CESC/UEMA, using specialised literature and/or by comparison, using exsiccates kept at HABIT and revised works on the family such as Arbo (1995, 2005) and Rocha et al. (2012). Records on the geographical distribution and ecological aspects of the species were obtained by consulting works compiled in the Lista de Espécies da Flora do Brasil (2016), Arbo (1995, 2005) and Rocha et al. (2012).

However, for the purposes of this research we chose to use the classification of Cronquist (1981), where the Turneraceae *sensu stricto* clade forms a family isolated from the Passifloraceae and Malesherbiaceae.

To confirm the identification of the specimens, duplicates of the specimens collected were sent to a taxonomist specialising in the Turneraceae family for identification at the specific level. Once identified, the specimens were duly organised in the LABIVE, serving both as testimonial material for databases and to support future research (Figure 5 C and D).

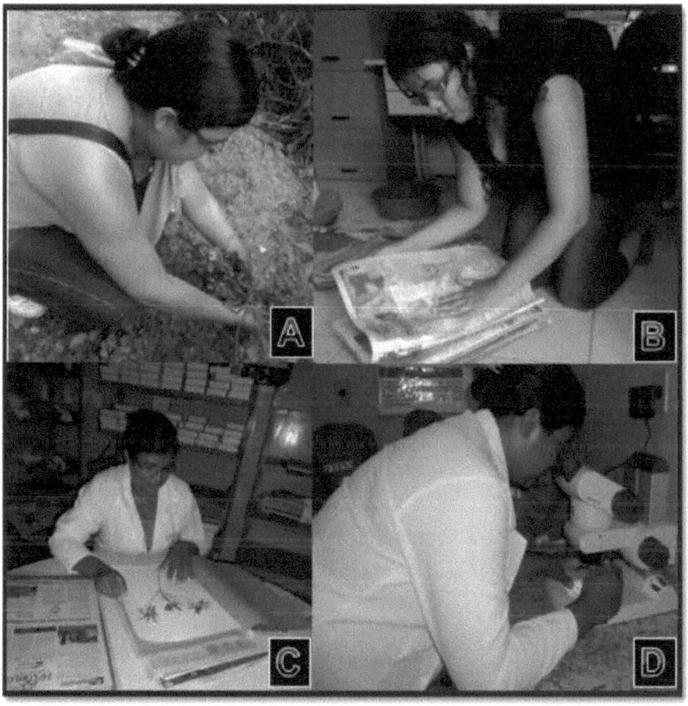

Fig. 5- A- Collection of botanical material; B- Botanical pressing; C- Botanical exsiccate; D- Identification of botanical material. Photo: SOUSA, D. A.

5. TURNERACEAE Juss. FROM APA-BURITI DO MEIO

Two genera (*Turnera* and *Piriqueta*) with seven species were identified for the Buriti do Meio APA. The *Turnera* genus had five species, *Turnera coerulea*, *T. melochioides*, *T. pumilea*, *T. scabra* and *T. subulata*, while the *Piriqueta* genus had two species, *Piriqueta duarteana* and *P. guianensis*. These species of Turneraceae were found in open environments, along roadsides and on trails within the vegetation, all of which have herb and/or subshrub habits (table 1).

Table 1- List of Turneraceae species collected in the Buriti do Meio APA, municipality of Caxias/MA.

Gender	Species	Habit	Environment
Piriqueta	*Piriqueta duarteana* (Cambess.) Urb.	Herbs/ Subarbuses	Open environment/Roadside and trails within vegetation
	Piriqueta guianensis N.E.Br.	Herbs	Open environment
Turnera	*Turnera coerulea* DC.	Herbs/ Subarbuses	Roadside verges and trails within the vegetation
	Turnera melochioides Cambess.	Herbs	Roadside verges and trails within the vegetation
	Turnera pumilea L.	Herbs/ Subarbuses	Open environment/Roadside and trails within vegetation
	Turnera scabra Millsp.	Herbs	Open environment
	Turnera subulata Sm.	Herbs/ Subarbuses	Open environment/Roadside and trails within vegetation

Source: Research data.

Compared to the work by Conceição et al. (2012), for the Inhamum APA, municipality of Caxias/MA, only the genus *Turnera* was recorded, with only one specimen identified at species level and one at genus level, and in the work by Nascimento; Conceição (2011), for the Olho d'água do Raposo Quilombola community, in the municipality of Caxias/MA, only one species was recorded, showing a greater specific richness for the Buriti do Meio APA when compared to the study carried out in the Inhamum APA and the Olho d'água do Raposo community.

Although only the species *P. duarteana* (Figure 5), *P. guianensis* (Figure 6), *T. coerulea* (Figure 7), *T. melochioides* (Figure 8), *T. pumilea* (Figure 9), *T. scabra* (Figure 10) and *T. subulata* (Figure 11) *were* collected for the Buriti do Meio APA, this study is relevant because it presents new records of the family for the municipality of Caxias and a species considered a new occurrence for Maranhão, *Piriqueta guianensis*.

According to the geographical distribution shown in Table 2, there is a predominance of species occurring in the Northeast region, predominantly in the states of Bahia, Ceará, Pernambuco and Piauí. As for Maranhão, the species *Piriqueta guianensis*, based on the Brazilian Flora List (2017), is a new record for the state.

According to Rocha et al. (2012), Turneraceae species occur preferentially in open environments and Caatinga and Restinga vegetation, while this study shows that they occur in typical Cerrado vegetation, but this may be associated in both studies with being in environments disturbed by anthropogenic action.
This also corroborates Arbo; Mazza (2011), since Turneraceae species can be distributed in cerrado, rupestrian grasslands and caatinga.

As well as the Turneraceae species showing a wide geographical distribution in the Northeast, it can also be seen that the

Turnera genus is used by the local population in a very important way in the Brazilian economy, since it is used for both ornamental and medicinal purposes. The leaves of *Turnera subulata,* for example, are used in folk medicine against amenorrhoea and are also used to treat tumours in the form of infusions (ARBO, 2005; BARBOSA et al., 2007).

Table 2- List of the geographical distribution of the Turneraceae family in the Buriti do Meio APA, municipality of Caxias/MA.

SPECIES	DISTRIBUCIÓN GEOGRAPHICAL	North-Northeast	Centre-west	Southeast
Piriqueta duartean	PA, TO AL, BA, CE, GO, MTMG	MA, PE, PI, RN,SE		
Piriqueta guianensis	RRAL	, BA CE,PB, PE, PI, RN, SE	,----	
Turnera coerulea	AM, AP PA, RR, TO	, BA, CE, MA, PE, PI, SE	GO, MS, MT	
Turnera melochioides	AM, AP, AL, BA, CE, GO, MS, PA, RO	, TOMA, PB, PE, PI, RN, SE	MT	

Turnera pumilea	*PA,* RR, TO	BA, CE, MA, PB, PE, PI, RN, SE	GO, MS, MT	MG, RJ
Turnera scabra	AC, AM, AP, PA, RR	AL, BA, CE, MA, PE, PI, RN	-----	ES
Turnera subulata	AM, AP, PA, RO, TO		DF, GO, MS, MT	ES, MG, RJ, SP
		AL, BA, CE, MA, PB, PE, PI, RN, SE		

Source: List of the Flora of Brazil, 2017.

After drawing up the taxonomic identification key and diagnosing each species collected in the study area, it was possible to see that the main characteristic for identifying the species was the colour of their petals, where for the genus *Piriqueta* it ranged from salmon to pink, while for the genus *Turnera* it ranged from white to yellow, along with the colour of their base. When differentiating the genera, the presence or absence of their glands was remarkable, as shown in Table 3.

Sampaio (2006) highlights the importance of the taxonomy of species, based on the characteristics of the flowers, especially the corolla, as a factor for categorising plants, as used by Tournefort (1656-1708).

Table 3_ List of the main morphological characteristics used to identify the Turneraceae species collected in the Buriti do Meio APA.

Species	Types of Flowers	Petal colouring	Colour of the petal base	Stipulas	Glands
P. duarteana	Heterostyl	Salmon, Orange	Yellow or with dark orange streaks	Absent	Absence of glands
P. guianensis	Heterostyl	Rose	Yellow	Absent	Absence of glands
T. coerulea	Heterostyl	Yellow	Dark yellow	Reduced to coléteres	Bigland
T. melochioides	Heterostyl	Yellow	Yellow	Inconspicuous	Bigland
T. pumilea	Homostila	White	Yellow with slight violet streaks	Absent	Bigland
T. subulata	Heterostyl	Yellow, White	Black bluish	Filiforms, Reduced to coléteres	Bigland
T. scabra	Heterostyl	Yellow	Dark yellow	Reduced to coléteres	Bigland

Source: Research data.

6. KEY FOR THE IDENTIFICATION OF GENUS AND SPECIES OF THE TURNERACEAE FAMILY FOR THE BURITI DO MEIO ENVIRONMENTAL PROTECTION AREA, CAXIAS/MA.

5. Absence of glands on the petiole, salmon to pink corolla *Piriqueta* (2)

5.1 Presence of glands on the petiole, white to yellow corolla *Turnera* (3)

6. Leaves with crenate margin .... *Piriqueta duarteana*

6.1 Leaves with wavy margin................................. *Piriqueta guianensis*

7. Homostylous flowers.. *Turnerapumilea*

7.1 ' ....Heterostylous flowers4

8. Yellow petals with a bluish base .... *Turnerasubulata*

8.1 ' Yellow petals with dark yellow base 5 *9.* membranaceous consistency.. *Turneracoerulea*

5.1' Consistency .... papyraceae ,,,,,6

10. Hirsute branches, elliptical leaf blade ....*Turnera melochioides*

6.1' Tomentose branches, ovate leaf blade ....

.... *Turnorascabra*

7. DESCRIPTION OF SPECIES OF THE TURNERACEAE FAMILY FOR THE BURITI DO MEIO ENVIRONMENTAL PROTECTION AREA, CAXIAS/MA.

7.1- *Piriqueta duarteana* (Cambess.) Urb. Jahrb. Konigl. Bot. Gart. Berlin 2: 66. 1883

Fig.6- ***Piriqueta duarteana.*** **Photo: SAMPAIO, N. A.**

Herb/sub-shrub with tomentose, decumbent branches, cylindrical and branched stem with a greenish hairy colour and simple and glandular trichomes. **Stipules** not visualised. **Petiole** 0.3-0.6 cm long, without glands. **Leaves** simple, alternate, 1.8-3.3 cm long x 1-1.7 cm wide, ovate, fleshy, membranous, margin crenate, apex acute to obtuse, base cuneate. **Peduncle** ca. 0.4 cm long. **Bracts** ca. 0.5 cm long, linear-lanceolate, with entire margin. Axillary **flowers** with radial symmetry, pentamerous and heterostylous. **Sepals** ca. 1.4 cm long x

0.4 cm wide, gamosepalous, green with yellow pigments, lanceolate with a smooth margin. **Petals** ca. 1.5 cm long x 0.6 cm wide, salmon to dark orange colour and yellow base or with dark orange streaks, entire margin. **Ovary** ca. 0.3 cm long, ovoid, tricarpellate gynoecium.

Material examined: MARANHÃO: Caxias, Buriti do Meio Environmental Protection Area: Cerrado. 21.I.2013, fl., *Sampaio, N. A. & Conceição, G. M., 03* (HABIT).

Synonym: *Turnera duarteana* Cambess.

Common name: *piriqueta*.

Geographic distribution: North: Pará and Tocantins; **Northeast:** Alagoas, Bahia, Ceará, Maranhão, Pernambuco, Piauí, Rio Grande do Norte and Sergipe; **Centre-West:** Goiás and Mato Grosso; **Southeast:** Minas Gerais.

Phytogeographic domain: Amazon, Caatinga, Cerrado and Atlantic Rainforest.

Comments: *Piriqueta duarteana* has a herbaceous or subshrubby habit and is commonly found in open environments, on roadsides and trails within vegetation.

7.2- *Piriqueta guianensis* N.E.Br. Trans. Linn. Soc. London, Bot. 6: 30. 1901

Fig.7- ***Piriqueta guianensis*. Photo: SAMPAIO, N. A.**

Herb with tomentose branches and cylindrical stem, greenish hairy colour and simple trichomes. **Stipules** not visible. **Petiole** 0.3-0.5 cm long, without glands. **Leaves** simple, alternate, 1-2 cm long x 0.5-1.4 cm wide, ovate, fleshy consistency, membranaceous, wavy margin, acute apex, cuneate base. **Peduncle**. 0.8 cm long. **Bract**. 0.5 cm long, linear-lanceolate, with entire margin. Axillary **flowers** with radial symmetry, pentamerous and heterostylous. **Sepals** ca. 0.9 cm long x 0.3 cm wide, green in colour, linear with a smooth margin. **Petals** ca. 1.2 cm long x 0.5 cm wide, pink in colour with a yellow base. **Ovary** ca. 0.4 cm long, ovoid, tricarpellate gynoecium.
Material examined: MARANHÃO: Caxias, Buriti do Meio Environmental Protection Area: Cerrado. 13.III.2013, fl., *Sampaio, N. A. & Conceição, G. M., 08* (HABIT).
Common name: *piriqueta*.

Geographic distribution: North: Roraima; **Northeast:** Alagoas, Bahia, Ceará,

Paraíba, Pernambuco, Piauí, Rio Grande do Norte and Sergipe.

Phytogeographic domain: Amazon, Caatinga, Cerrado and Atlantic Rainforest.

Comments: *Piriqueta guianensis* has a herbaceous habit, commonly found in open environments.

7.3- *Turnera coerulea* DC. Prodr. 3: 346. 1828

Fig.8- ***Turnera coerulea.*** **Photo: SAMPAIO, N. A.**

Herb/subshrub with a cylindrical, hairy, greenish stem and simple trichomes. **Stipules** reduced to collettes. **Petiole** 0.4 cm long, with 2 glands. Simple alternate **leaves**, elliptic, 12.2 cm long x 0.4-0.5 cm wide, ovate, fleshy consistency, membranous, cuneate base, acute to obtuse apex, serrate margin. **Peduncle** ca. 0.4 cm long. **Bracts** ca. 0.4 cm long, linear, hairy with entire margin. Axillary **flowers** with radial symmetry, pentamerous and heterostylous. **Sepals** ca. 0.9 cm long x 0.3 cm wide, green with yellow pigmentation, linear with entire margin. **Petals** ca. 0.5 cm long x 0.3 cm wide, gamopetalous, yellow in colour with a dark yellow base and entire margin. **Ovary** ca. 0.2 cm long, ovoid.

Material examined: MARANHÃO: Caxias, Buriti do Meio Environmental Protection Area: Cerrado. 24.I.2013, fl., *Sampaio, N. A. & Conceição, G. M., 05* (HABIT).

Common name: chanana.

Geographic distribution: North: Amazonas, Amapá, Pará, Roraima and Tocantins; **Northeast:** Bahia, Ceará, Maranhão, Pernambuco, Piauí and Sergipe; **Centre-West:** Goiás, Mato Grosso do Sul and Mato Grosso.

Phytogeographic domain: Amazon, Caatinga, Cerrado, Atlantic Rainforest and Pantanal.

Comments: *Turnera coerulae* has a herbaceous or sub-shrub habit and is commonly found on roadsides and trails within vegetation.

7.4- - *Turnera melochioides* Cambess. Fl. Bras. Merid. (4th ed.) 2(16): 219. 1829 [1830]

Fig.9- ***Turnera melochioides*****. Photo: SAMPAIO, N. A.**

Hirsute **herbs** with cylindrical, branched stems, a greenish hairy colour and simple, smooth or wavy trichomes. **Stipules** inconspicuous. **Petiole** 0.2-0.4 cm long, with 2 glands. **Leaves** simple, alternate helical, elliptic, lanceolate, 1-2.2 cm long x 0.4-0.6 cm wide, fleshy consistency, papery, obtuse apex, cuneate base, serrate margin. **Peduncle** ca. 0.2 cm long. **Bracts** ca. 0.3 cm long, linear, hairy with entire margin. Axillary **flowers** with radial symmetry, pentamerous and heterostylous. **Sepals** ca. 1 cm long x 0.4 cm wide, green in colour, linear with entire margin. **Petals** ca. 0.6 cm long x 0.4 cm wide, gamopetalous, yellow to the base and entire margin. **Ovary** ca. 0.4 cm long, ovoid, tricarpellate gynoecium.
Material examined: MARANHÃO: Caxias, Área de Proteção Ambiental do Buriti do Meio: Cerrado. 22.I.2013, fl., *Sampaio, N. A. & Conceição, G. M., 07* (HABIT).
Common name: chanana, star plucker.

Geographic distribution: North: Amazonas, Amapá, Pará, Rondônia and Tocantins; **Northeast:** Alagoas, Bahia, Ceará, Maranhão, Paraíba, Pernambuco, Piauí, Rio Grande do Norte and Sergipe; **Centre-West:** Goiás, Mato Grosso do Sul and Mato Grosso; **Southeast:** Minas Gerais.

Phytogeographic domain: Amazon, Caatinga, Cerrado and Atlantic Rainforest. **Comments:** *Turnera melochioides* has a herbaceous habit and is commonly found on roadsides and trails within vegetation.

7.5- *Turnera pumilea* L. Syst. Nat. (ed. 10) 2: 965. 1759

Fig. 10- *Turnera pumilea*. Photo: SAMPAIO, N. A.

Herb/sub-shrub with a cylindrical, branched stem, greenish colour with brown pigmentation and simple, wavy trichomes. **Stipules** not visible. **Petiole** 0.5-0.6 cm long, with 2 glands in the lower third. Simple alternate **leaves**, lanceolate 1.6-2.5 cm long x 0.4-1.5 cm wide, fleshy consistency, membranaceous, ovate to elliptic, margin repanda (near the base of the leaf), serrate, apex acute, base attenuated and nervation peninervate. **Peduncle** ca. 0.6 cm long. **Bracts** ca. 0.6 cm long, linear-lanceolate, hairy with entire margin. Axillary **flowers** with radial symmetry, pentamerous and homostylous. **Sepals** ca. 0.6 cm long x 0.4 cm wide, green in colour, gamosepal, linear with a smooth margin. **Petals** ca. 0.9 cm long x 0.5 cm wide, dialipetalous, white in colour with a yellow base and sometimes showing light violet streaks and an entire margin. **Ovary** ca. 0.2 cm long, ovoid and globose.

Material examined: MARANHÃO: Caxias, Buriti do Meio Environmental Protection Area: Cerrado. 13.XII.2012, fl., *Sampaio, N.*

A. & Conceição, G. M., 01 (HABIT).
Common names: albina, chanana.

Geographic distribution: North: Pará, Roraima and Tocantins; **Northeast:** Bahia, Ceará, Maranhão, Paraíba, Pernambuco, Piauí, Rio Grande do Norte and Sergipe; **Centre-West:** Goiás, Mato Grosso do Sul and Mato Grosso; **Southeast:** Minas Gerais and Rio de Janeiro.
Phytogeographic domain: Amazon, Caatinga, Cerrado and Atlantic Rainforest.
Comments: *Turnera pumilae* has a herbaceous or sub-shrub habit and is commonly found in open environments, on roadsides and trails within vegetation.

7.6- *Turnera scabra* Millsp.Publ. Field Columb. Mus., Bot. Ser. 2(1): 77. 1900

Fig.11- *Turnera soabra*. Photo: SAMPAIO, N. A.

Herb with tomentose branches and cylindrical stem, greenish colour and simple trichomes. **Stipules** reduced to collettes. **Petiole** ca. 0.3 cm long, with 2 glands in the lower third. **Leaves** simple, alternate,

1.6-2.5 cm long x 0.4-1.5 cm wide, ovate, fleshy, papery, serrate margin, acute apex, attenuated base. **Peduncle** ca. 0.3 cm long. **Bracts** ca. 0.4 cm long, linear, hairy with entire margin. Axillary **flowers** with radial symmetry, pentamerous and heterostylous. **Sepals** ca. 08 cm long x 0.3 cm wide, green with yellow pigmentation, linear with entire margin. **Petals** ca. 1.3 cm long x 0.5 cm wide, yellow colour and dark yellow base, entire margin. **Ovary** 0.3 cm long, ovoid.
Material examined: MARANHÃO: Caxias, Área de Proteção Ambiental do Buriti do Meio: Cerrado. 22.I.2013, fl., *Sampaio, N. A. & Conceição, G. M., 06* (HABIT).
Synonym: *Turnera ulmifolia* var. *intermedia* Urb.

Common name: chanana.
Geographic distribution: North: Acre, Amazonas, Amapá, Pará and Roraima; **Northeast:** Alagoas, Bahia, Ceará, Maranhão, Pernambuco, Piauí and Rio Grande do Norte; **Southeast:** Espírito Santo.
Phytogeographic domain: Amazon, Caatinga and Atlantic Rainforest.
Comments: *Turnera scabra* has a herbaceous habit and is commonly found in open environments.

7.7- *Turnera subulata* Sm.The Cyclopaedia; or, universial dictionary of arts, 36: n. 2. 1817.

Fig.12- *Turnera subulata.* Photo: SAMPAIO, N. A.

Erect/decumbent **herb/sub-shrub** with a cylindrical, woody stem,

greenish colour and strigose to pubescent, smooth or wavy trichomes. **Stipules** reduced to collettes. **Petiole** 0.6-1.9 cm long, with 2 glands in the lower third. **Leaves** simple, obovate, elliptic, alternate helical, 1.7-6.8 cm long x 0.7-3.5 cm wide, fleshy consistency, membranaceous, margin repanda (near the base of the leaf), serrate- crenate, apex acute to obtuse, base attenuated and nervation peninereve. **Peduncle** ca. 0.6 cm long. **Bracts** ca. 0.7 cm long, linear-lanceolate, with entire margin. Axillary **flowers** with radial symmetry, pentamerous and heterostylous. **Sepals** ca. 1.2 cm long x 0.4 cm wide, green with yellow pigmentation, linear with entire margin. **Petals** ca. 1.2 cm long x 0.6 cm wide, yellow/white in colour with a bluish-black base and entire margin. **Ovary** ca. 0.3 cm long, ovoid, unicarpellate gynoecium.
Material examined: MARANHÃO: Caxias, Buriti do Meio Environmental Protection Area: Cerrado. 24.I.2013, fl., *Sampaio, N. A. & Conceição, G. M., 04* (HABIT).

Synonym: *Turnera ulmifolia* var. *elegans* (Otto) Urb.
Common names: chanana, passion flower.

Geographic distribution: North: Amazonas, Amapá, Pará, Rondônia and Tocantins; **Northeast:** Alagoas, Bahia, Ceará, Maranhão, Paraíba, Pernambuco, Piauí, Rio Grande do Norte and Sergipe; **Centre-West:** Distrito Federal, Goiás, Mato Grosso do Sul and Mato Grosso; **Southeast:** Espírito Santo, Minas Gerais, Rio de Janeiro and São Paulo.

Phytogeographic domain: Amazon, Caatinga, Cerrado and Atlantic Rainforest.

Comments: *Turnera subulata* has a herbaceous or subshrubby habit and is commonly found in open environments, along roadsides and trails within vegetation

8. CONCLUSION

This study provides the first data on the Turneraceae family for the Buriti do Meio Environmental Protection Area, and is therefore the first record of species for the area. However, the data obtained does not total the number of existing species for the Buriti do Meio Environmental Protection Area, which is of the utmost importance for getting to know the local flora, adding new records of the family for Maranhão.

In view of the great importance of studies on the Turneraceae family, it can be seen that there are still few studies on Maranhão, a fact that can be confirmed by the new record of the species *Piriqueta guianensis* for the state. In this way, the research has made a significant contribution to the inventory of the first records of Turneraceae for the Buriti do Meio APA and, consequently, to expanding the taxonomic and floristic knowledge of the family for the state of Maranhão.

SUMMARY

The Turneraceae Kunth ex DC. family comprises herbs to shrubs with alternate leaves usually bearing petiolate and marginal extrafloral nectaries, homostylous or heterostylous flowers, with or without a corona. The family has around 12 genera and 226 species, distributed throughout the tropical and subtropical regions of the world, with America and Africa as its centre of diversity. The aim of this study was to carry out a taxonomic study of the species of the Turneraceae family that occur in the Buriti do Meio Environmental Protection Area, in the municipality of Caxias/MA. The Buriti do Meio APA is located between the geographical coordinates -04° 54' 48.1" South latitude and -043° 06' 49.2" West longitude, with typical cerrado vegetation. Specimens were collected from the study area and photographed. After herborisation and identification of the botanical material, exsiccates were added to the collection of the Aluízio Bittencourt Herbarium at CESC/UEMA. The research presented two genera (*Piriqueta* and *Turnera*), distributed in seven species: *Piriqueta duarteana, P. guianensis, Turnera coerulea, T. melochioides, T. pumilea, T. scabra* and *T. subulata*. The Turneraceae species inventoried for the Buriti do Meio Environmental Protection Area provide the first records of the family for the study area, and are of paramount importance for understanding the local flora and recording a new occurrence for Maranhão: *Piriqueta guianensis*.

Keywords: Plant diversity; *Piriqueta*; *Turnera*

REFERENCES

ANTÔNIO, M. A. General Pharmacological Actions of *Turnera ulmifolia* L. on the Inflammatory Response. **Master's dissertation.** 1996. State University of Campinas, São Paulo. 1996.

ANTÔNIO, M. A. & SOUZA, A. R. M. Oral anti-inflammatory andanti-ulcerogenic activities of a hydroalcoholic extractand partition Ed fractions of *Turnera ulmifolia* (Turneraceae). **J. Ethnopharmacol,** v.61, p. 215-228, 1998.

APG III. An update of the Angiosperm Phylogeny Group classification for the orders and families of flowering plants: APG. **Botanical Journal of the Linnean Society**. v. 161, p. 105-121, 2009.

ARBO, M. M. **Turneraceae Part I. *Piriqueta*.**Flora Neotropical, v. 67, p. 124128, 1995.

ARBO, M. M. **Systematic studies on *Turnera* (Turneraceae).** I Series *Salicifoliae y Stenodictyae*. Bonplandia, v. 9, p. 151-208, 1997.

ARBO, M. M. **Systematic studies on *Turnera* (Turneraceae).** II Series *Annulares, Capitatae, Microphyllae y Papiliferae. Bonplandia,* v.10, p. 1-82, 2000.

ARBO, M. M. **Systematic studies on *Turnera* (Turneraceae).** III Series *AnomalaeyTurnera. Bonplandia*, v. 14, p. 115-318, 2005.

ARBO, M. M. **Systematic studies on *Turnera* (Turneraceae).** IV. Series *Leiocarpae, Conciliatae* y *Sessilifoliae* Bonplandia, v.17, n. 2, p. 107-334, 2008.

ARBO, M. M. *Turneraceae* in **Lista de Espécies da Flora do Brasil**. Botanical Garden of Rio de Janeiro. 2014. Available at: <http://floradobrasil.jbrj.gov.br/jabot/floradobrasil/FB240>. Accessed on 28 February 2017.

ARBO, M.M.; MAZZA, S.M. **The major diversity centre for neotropical Turneraceae** .Systematics and Biodiversity, v. 9, p. 203-210, 2011.

BARBOSA, D. J.; SILVA, K. N. & AGRA, M. F. Comparative Pharmacobotanical Study of the Leaves of *Turnera chamaedrifolia* Cambess. and *Turnera subulata* Sm. (Turneraceae). **Revista Brasileira de farmacognosia.** v. 17, n. 3, p. 396-413, 2007.

BATISTA, R. W. Antioxidant Effects (*in vitro* and *in vivo*) of Flavonoid-Rich Species Belonging to the Genus *Turnera*, Family Turneraceae. **Master's dissertation.** 2005. State University of Campinas, Campinas, São Paulo. 2005.

BRAZIL. **Ministry of Cities** 2006. Available at: <http://www.cidades.gov.br/imagens/stories/arquivosSNPURedeAvaliação/C axiasanexo02MA.pdf> Accessed on 10 January 2014.

BRASIL. **Curso de Identificação Botânica de Espécies Arbóreas**

da Região Amazônica. Brasília: Editora Rima. 2009.

BRITO FILHO, S. G. Pheophytins and Glycolysed Steroids from *Turnera subulata* Sm. (Turneraceae). **Master's dissertation.** 2011. Federal University of Paraíba. João Pessoa, Paraíba. 2011.

CAMARGO, E. E. S.; VILEGAS, W. **Quality control of polar extracts of *Turnera difusa* Willd. Ex Schult., Turneraceae.** Revista Brasileira de Farmacognosia:v. 20, n. 2, p. 228-232, 2008.

CONCEIÇÃO, G. M.; RUGGIERI, A. C.; SILVA, E. O.; NUNES, C. S.; GALZERANO, L.; NERES, L. P. **Phanerogamic flora of the Inhamum Municipal Environmental Protection Area, Caxias/MA, Brazil.** *In:* Biodiversity in the Inhamum Environmental Protection Area. São Luís: UEMA. 2012.

CRONQUIST, A.**An integrated system of classification of flowering plants.**The New York Botanical Garden, Columbia University Press, New York.1262p. 1981.

CRONQUIST, A.**The evolution and classification of Flowering Plants.** 2ª ed. New York: The New York Botanical Garden, 555p. 1988.

HARLOW, W. M.; HARRAR, E. S.; HARDIN, J. W.; WHITE, F. M. **Testbook of dendrology: Covering the important forest trees of the United states and Canada.**Singapov: McGraw-Hill, Inc. 7th ed., 501p. 1991.

KAMPF, A. N.; COSTA, G. J. C. **Commercial production of ornamental plants.** Guaíba: Agropecuária, 2000.

LISTA DE ESPÉCIES DA FLORA DO BRASIL. **Botanical Garden of Rio de Janeiro**. Available at: <http://floradobrasil.jbrj.gov.br/>. 2017.

MARX, R. B. **Art and landscape: selected lectures.** José Tabacow, organisation and commentary. 2nd ed. São Paulo: Studio Nobel, 2004.

MENDONÇA, R. C., FELFILI, J. M.; WATER, B. M. T., SILVA JÚNIOR, M. C., REZENDE, A. V., FILGUEIRAS, T. S. & NOGUEIRA, P. E. & FAGG, C. W. **Vascular flora of the Cerrado.** *In:* Sano, S. M. & Almeida, S. P. Cerrado: Environment and flora. Planaltina: EMBRAPA-CPAC. p. 307-556. 2007.

NASCIMENTO, J. M. & CONCEIÇÃO, G. M. Plantas Medicinais e Indicações Terapêuticas da Comunidade Quilombola Olho D'água do Raposo, Caxias, Maranhão, Brazil. **Journal of Biology and Pharmacy**, v. 6, n. 2, 2011.

NATIONAL INFORMATION STANDARDS ORGANISATION (2005). ANSI/NISOZ39.19-2003: guidelines for the construction, format, and management of monolingualthesauri. 2005. Available at: <http://www.niso.org/standards/resources/Z39 19-2005.pdf#search=%22z39. 19%22. Accessed on 28 February 2017.

NEFFA, V. G. S. & FERNÁNDEZ, A. **Chromosome Studies in *Turnera* (Turneraceae).** Genetics and Molecular Biology.v. 23, n. 4, p.925-930, 2000.

ROCHA, L. N. G.; MELO, J. I. M. & CAMACHO, R. G. V. Flora of Rio Grande do Norte, Brazil: Turneraceae Kunth ex DC. **Revista Rodriguésia,** v. 63, n. 4, 2012.

SAMPAIO, G. **Botany** (Chap. 3). 2006. Available at: <http://www.fc.up.pt/pessoas/jpcabral/index_felex/GS_brocheera.pdf>
Accessed on 28 February 2017.

SILVA, P. H. S.; BEZERRA, T. T.; SANTOS, S. R. R.; ALMEIDA, N. M.;
SIQUEIRAFILHO, J. A. Pollen Flow and Isopletia in *Piriqueta duarteana* Var.

Ulei (Turneraceae) a dystyllic species. 64th National Botanical Congress Belo Horizonte (10-15 Nov 2013), 2013.

SILVA, .F.N. Hydrographic map and location of the study area, collection points, Buriti do Meio, Caxias/MA. Google Earth Pro. ArcGis 9.3. Consulted on 22/01/2016. Werton Francisco Nobre Silva. 2016.

STEVENS, P. F. Angiosperm phylogeny website. http://www.mobot.org./MOBOT/research/Apweb/. 2001. Accessed on 10 February 2017.

SUBRAHMANYAM, N. S. **Modern plant taxonomy.** New Delhi: Vikas Publishing House PVT Ltd. 494p. 1995.

THULIN, M.; RAZAFIMANDIMBISON, S. G.; CHAFE, P.; HEIDARI, N. KOOL, A. & SHORE, J. S. **Phylogeny of the Turneraceae clade (Passifloraceaes.*l.*): Trans- Atlantic disjunctions and two new genera in Africa.** Taxon,v. 61, p. 308-323, 2012.

URBAN, I. **Monographie des familie Turneraceen.** Vol 2. Jahrb. Konigl. Bol. Gart. Berlin, p. 1-152, 1883.

WEBERLING, F.; SCHWANTES, H. **Taxonomia Vegetal.** São Paulo: ed. Pedagógica e Universitária. 314p. 1986.

ZAMBERLAN, P. M. **Phylogeny of *Passiflora* L. (Passifloraceae): infra-subgeneric questions.** Master's thesis. 2007. UFRGS. Porto Alegre. 2007.

Table of contents

MIX
Papier aus verantwortungsvollen Quellen
Paper from responsible sources
FSC® C105338

Printed by Books on Demand GmbH, Norderstedt / Germany